H C Q

The New Highway Code

by

BRIAN M STRATTON

Department of Transport ADI

First Time (Driving)

Telephone (0860) 260720

HCQA

The New Highway Code Questions & Answers

by

BRIAN M STRATTON
Department of Transport   ADI

The title, layout and contents of this book are copyright
© Brian M Stratton 1993

The Highway Code extracts are reproduced with the permission of the Controller of Her Majesty's Stationery Office.

No part of this book may be reproduced, stored in a retrieval system or transmitted in any form or by any means, electronic, electro static, magnetic tape, mechanical, photocopying, recording or otherwise without permission in writing from the publisher.

Note: Whilst every effort has been made to ensure the complete accuracy of this book, neither the author nor publisher can accept liability for any error or misinterpretation of the information contained therein.

First published 1993
Reprinted February 1993

ISBN : 0 9514415- 8 - 2

Printed in Great Britain

Published by : First Time (Driving)

Telephone: (0860) 260720

Copyright Brian M Stratton 1993

ABOUT THE AUTHOR

The author of this book is a Department of Transport Approved Driving Instructor and also holds the following qualifications:

DIAmond Advanced Instructor

The Diploma in Driving Instruction

Cardington 'Special' Driving Test Grade A

RoSPA Diploma

Member of IAM, DIA, MSA, GEM

ADITE No 54
(Approved Driving Instructor Training Establishment)

Check Tested Grade 6
(Highest Grade)

He has attended and successfully completed a Police Better Driving Course (including skid-pan training) and several Rally Driving courses.

Having successfully taught many learner drivers to pass first time, and by 'sitting in' on many driving tests, he has gained considerable knowledge of both the test itself and the examiner's requirements.

Also by the same author:

The Driving Test: *Graphic Traffic Version*

Hill Start Blues : A New Driving Manual for the 1990s
(Revised Edition September 1991)

ADI Part III : Essential Information
(Revised Edition July 1992)

ADI Part III and Instructor's Guide

'L' on Wheels

[For more details of these titles, see pages at end of this book]

First Time (Driving)

Tel: (0860) 260720

FOREWORD

- Use this book in conjunction with the Highway Code. By completing the questions in the A-Z Modules, p 25, you will find that your attention is focussed on all aspects of the code.

- This book will also help to prepare for the way in which questions are phrased on the driving test, and how to give your answers.

- Having passed your test, always keep a copy of the Highway Code in your car and update it as each new edition is brought out.

Look upon the Highway Code as something you will refer to throughout your driving career, NOT just on your driving test.

CONTENTS

About the Author

Also by the same Author

    ADI Part III   -   Essential Information

    The Driving Test   -   Graphic Traffic Version

    Hill Start Blues   -   A New Driving Manual for the 1990s

    ADI Part III and Instructor's Guide

Foreword

		Page No
Section 1	What is the Highway Code?	9 - 10
Section 2	Why a New Edition?	11 - 12
Section 3	Changes to the Code	13 - 14
Section 4	How are the Questions Chosen?	15 - 16
Section 5	How will the Examiner Phrase Questions?	17 - 18
Section 6	How should Answers be Given?	19 - 20
Section 7	No Excuses ...	21 - 22
Section 8	How to Study with this Book	23 - 24
Section 9	Questions & Answers in Modules A - Z	25 - 78
Section 10	Traffic Signs	79 - 84
Section 11	Answers : Traffic Signs	85 - 88
Section 12	And Finally	89
	Driver Training Standards	90 - 91

No part of this book may be reproduced, stored in a retrieval system or transmitted in any form or by any means, electronic, electrostatic, magnetic tape, mechanical, photo-copying, recording or otherwise without permission in writing from the publisher.

SECTION 1

WHAT IS THE HIGHWAY CODE ?

The Highway Code is a mixture of law and good advice.

Breaking one of the rules of the highway code is not necessarily punishable, but it is likely that such a transgression is also contravening some specific rule in the Road Traffic Act, and a road user could be liable for prosecution for that.

ADVERTS

The very first Highway Code appeared in 1931, cost one old penny, was 21 pages long and about half the size of this book. It also carried adverts for, amongst other things, car batteries, insurance and motor oil.

This Code gave motorists a summary of features of the 1930 Road Traffic Act which governed the use of a motorcar on the road. It was issued as a code of directions for the guidance of road users.

HORSE AND CARRIAGE

A supplementary Road Traffic Act of 1934 prompted a further edition of the Highway Code. Still costing one penny this edition had a blue cover and no longer carried adverts, featured line drawings to illustrate various aspects of motoring, including giving signals with your whip, if you were the driver of a horse-drawn vehicle!
This edition featured, for the first time, traffic signs. It also introduced the new 30 mph speed limit in built-up areas.

DINKY TOYS

The Highway Code has had nine editions, each one reflecting changes in road / traffic conditions, and additions to the Road Traffic Act.

Diagrams were improved from the original drawings and by the 1968 edition (costing three old pence), Dinky Toys (small model cars) were used on a painted background to illustrate lane disciplines, roundabout approach etc. In the 1990s these illustrations were brought up to date with the introduction of colour computer graphics.

NOT JUST FOR L - DRIVERS

Although most people associate the Highway Code with learning to drive, it is designed for all road users - pedestrians, cyclists/motor-cyclists, people in charge of animals - to advise them of the safe and correct way to proceed.
It is therefore in all road users' own interest to keep themselves aware of new editions and new guidelines/laws.

SECTION 2

WHY A NEW EDITION ?

The Highway Code was last fully revised in 1978. Minor changes have been made since then but nothing major, until now . . . The style, format and layout have been improved to make it easier to read, along with revised illustrations. It is also hoped that road users will be encouraged to use it on a regular basis.

RECENT DEVELOPMENTS

This major revision of the Highway Code takes into account recent developments in road traffic law and also a need to up-date information and provide a guide which will benefit all road users.

CLEAR AND SIMPLE

The law's demands (a section found at the back of the old Highway Code) has been clarified and provides a simple guide to the law. For the precise wording of the law the various acts and regulations can be consulted at most main reference libraries.

SECTION 3

CHANGES TO THE CODE

With the new edition, new rules have been introduced which are listed below.
Virtually all of the code has been changed with rules being re-written to make them easier to read or understand; more 'user-friendly'.

The only rule left unchanged is Rule 67 (formerly 59) which states : 'Be careful near a parked ice cream van as children are more interested in ice cream than in traffic.'

THE NEW RULES

The rule number is given on the left and then the subject of the rule.

PEDESTRIANS

16	Use of Puffin crossings.
20	Use of textured paving at pedestrian crossings to assist blind or partially sighted pedestrians.

DRIVERS, MOTORCYCLISTS AND CYCLISTS

36-37	New requirements for supervision of learner drivers and compulsory basic training for motorcyclists.
55	Reducing speed in residential areas.
59-62	Advice on driving in winter conditions.
77	Significance of flashing amber warning lights.
78	Explanation of police procedures for stopping vehicles

81	Hand signals given by horse riders.
88	Use of 'crawler' lanes by slow moving vehicles.
94	Lane discipline on dual carriageways.
116	Taking care at inoperative traffic lights.
147-148	Taking care at road works.

MOTORWAYS

159	Use of slip roads on access to motorway.
165	Lane discipline at motorway exits.

CYCLISTS

190	Advice on safety equipment.
195-197, 199, 204, 206-208 & 211	Advice on cycling safety.

ANIMALS

217-218, 221 & 223-224	New advice for horse riders.

TRAMWAYS

235-242	New section giving advice for road users on safety in proximity to tram operations.

SECTION 4

HOW ARE THE QUESTIONS CHOSEN ?

When the examiners are undergoing training, they are required to prepare relevant questions on the Highway Code and other motoring matters.
When examiners are carrying out the job on a daily basis they will also be required to think up new questions, particularly when new laws are introduced which affect road users.

OTHER MOTORING MATTERS

Other motoring matters covers a wide variety of motoring aspects. Questions could be asked on the cause of skids, what causes accidents, basic mechanics etc, etc.

TRICK QUESTIONS?

No ! There are no trick questions and the examiner will not set out to mis-lead you or make things difficult for you. However, it is possible that in the heat of the moment candidates do sometimes mis-hear questions. After the test, when asked to repeat them, these questions often become distorted to a certain degree as the following examples show :

One candidate thought the chances of her passing were remote when the examiner said "I'd like to put some questions to you on the Highway Code, not that it matters . . . "

What he'd actually said was " . . . on the Highway Code and other motoring matters"

Another candidate thought he had been asked "What is Rule No 2 ?"

The examiner had in fact asked "What is the two-second rule?"

So, no trick questions, but listen carefully to what the examiner is saying, and if you're not sure, or you mis-heard, ask him to repeat the question.

SECTION 5

HOW WILL THE EXAMINER ASK QUESTIONS ?

The questions are asked in a very straightforward manner.

You will be asked the questions at the end of the practical part of the driving test. When you return to the test centre the examiner will ask you to pull up, and will say to you "You may now stop the engine and release the seat-belt. I'd like to put some questions to you on the Highway Code and other motoring matters".

CONCISE

Questions are phrased clearly and concisely and in a way that can be easily understood.

An example : "What is the right hand lane of a motorway used for?"

When the question has been asked the examiner will then wait for your reply. If you say nothing the examiner will ask another question. When you give your answer the examiner will not say if it is correct or not, but will say "I see", or "Thank you" and then move on to the next question.

SUPPLEMENTARY QUESTIONS

If the examiner wanted more information from you regarding a certain topic, he may ask further questions following on from his original question.

An example : "When should you use your mirrors?"

Supplementary questions : "Why is it so important to check your mirrors?"

: "What will you know if you check your mirrors?"

: "If there was someone very close behind you, how would that affect your braking?"

IDENTIFYING TRAFFIC SIGNS

After the examiner has asked you questions, he will then ask you to identify some traffic signs. These will be chosen at random from his flip chart of signs. After your answer, or pause long enough for an answer, he will flick through to another sign. He will not say whether your answers are correct or not.

SECTION 6

HOW SHOULD ANSWERS BE GIVEN ?

When giving answers to the examiner's questions, be concise, clear and straightforward.

Don't launch into lengthy explanations. Keep your answer simple and to the point.

An example could be : "What must you always do before changing direction ?"

The answer is given simply as "Check the mirrors".

COMMON SENSE

If you don't know the answer to a particular question use your common sense and have an educated guess. The examiner would prefer you to use your initiative and think about it and give some sort of an answer rather than just saying "Dunno" without making any attempt at an answer. After all, when you are a qualified driver you will have to work things out for yourself - so give yourself a good start to your driving career and study the code and apply it in practice.

TRAFFIC SIGNS

When the examiner shows you traffic signs to identify you can use the shape to help you to remember what type of sign it is, ie warning, prohibitory, information etc. Having decided on that, the symbol within is usually self-explanatory.

If the examiner feels that your answer is not quite sufficient he may well ask you a supplementary question to confirm your understanding of a particular sign.

SECTION 7

NO EXCUSES ...

There really is no excuse for not studying the Highway Code. It's widely available, cheap to buy, easy to carry around and, because it's divided up into rules, it's easy to dip into and read a bit at a time.

Ideally, you should start studying the Highway Code as soon as you start learning to drive. It is far better to study for 15 - 20 minutes per day over a few months rather than cram it all in the few days before the test - if you try to cram you'll find yourself looking at the pages without any of it going in.

EXCUSES ... EXCUSES

ADIs (Approved Driving Instructors) across the country are used to hearing various 'explanations' as to why pupils have not studied the code.

Here is a list of excuses which have been given for not studying the Highway Code. All ADIs will have heard some or all of them.

1. I didn't have time .....
2. I forgot
3. I lost it
4. The shops had sold out (What, all of them ?)
5. I'll do it next week
6. I'll definitely do it next week
7. I promise I'll definitely do it the night before my test
8. It went through the washing machine
9. The dog ate it
10. On holiday, it fell into the swimming pool

Of course, you'd never use any of the above would you ... ?

SECTION 8

HOW TO STUDY WITH THIS BOOK

Use the book to 'dip' into as and when you have time to spare; perhaps on a bus or train journey - or while waiting to start such a journey.

Use the Record Sheet to record your score and the date.

Work towards achieving 100% in all modules. Those modules that you do not consistently score highly on should be studied in more depth.

NOT ONLY BUT ALSO

Obviously this book should be used in conjunction with the Highway Code. Your ADI should advise you as to the sections to study which are relevant to your particular stage of development or pertinent to forthcoming lessons.

It is important that you know not only the answers but the underlying principle and spirit behind the rules of the Code.

WRITE IT DOWN

If there is a particular rule or section that you are having difficulty in remembering, it can be helpful to write it down. The act of writing something out could help you to retain the fact(s) in your memory.

USE THIS FORM TO RECORD YOUR SCORE

MODULE	Score	:	Date	Score	:	Date	Score	:	Date	Score	:	Date
A												
B												
C												
D												
E												
F												
G												
H												
I												
J												
K												
L												
M												
N												
O												
P												
Q												
R												
S												
T												
U												
V												
W												
X												
Y												
Z												
Total												
÷ 26												
Average score												

SECTION 9

QUESTIONS MODULE A

A-1 If you are taking medicines, who should you ask if it is safe for you to drive?

A-2 What is the law regarding supervising drivers who accompany learner car drivers?

A-3 You want to learn to ride a motorcycle, scooter or moped, what must you do before riding on the road?

A-4 By law, what must be displayed when the vehicle is being driven by a learner?

A-5 When a vehicle is being driven by a qualified driver, 'L' plates must be covered or removed. Which vehicles are exempt from this rule?

* * * * *

A-6 What is the legally permitted alcohol level in blood, breath, urine, for drivers?

A-7 What does the Highway Code advise regarding fitness to drive after drinking at lunchtime or in the evening?

A-8 Whose responsibility is it for passengers under 14 to make sure seat belts are used?

A-9 If passengers are over 14, who is responsible for making sure that seat belts are used?

A-10 From 1 January 1993, children aged 1, 2 and 3, in the front passenger seat, must be restrained by what?

ANSWERS MODULE A

A-1 *If they are prescribed, ask the doctor; if OTC (over the counter) ask the pharmacist.*

A-2 *The supervisor must be aged at least 21, and have held (and currently hold) a full licence (British/exchangeable) for 3 years.*

A-3 *Take basic training with an approved body.*

A-4 *'L' plates must be displayed.*

A-5 *Driving school vehicles.*

* * * *

A-6 *Blood 80 mg/100 ml, Breath 35 microgrammes/100 ml, Urine 107 mg/100ml.*

A-7 *You may still be unfit to drive after drinking at lunchtime, or in the morning after drinking the previous night.*

A-8 *The driver's responsibility.*

A-9 *The passenger's responsibility.*

A-10 *A child seat or harness appropriate to the child's weight.*

QUESTIONS MODULE B

B-1 If you have children in the car, what does the Highway Code advise that you should do?

B-2 You are driving a car fitted with a phone. It starts to ring. What should you do?

B-3 You need to give signals to help and warn others - who?

B-4 Before moving off, what is the final check you should make?

B-5 Drivers should do what before carrying out any manoeuvre?

* * * * *

B-6 Before carrying out any manoeuvre a motorcyclist should do what?

B-7 How often should you check your mirrors?

B-8 What does the MSM sequence mean?

B-9 As a driver you should keep a special watch out for cyclists and motorcyclists. Why is this?

B-10 If you are driving a large or slow-moving vehicle, what could you do to help other road users?

ANSWERS MODULE B

B-1 Keep them under control.

B-2 Find a safe place to stop and answer it. If it is a hands-free phone you can answer it but keep your full attention on the road.

B-3 Other drivers, cyclists, horse riders, pedestrians.

B-4 Blind spot check - over your shoulder(s).

B-5 Check the mirrors.

* * * * *

B-6 Look behind. If, as a driver you see a motor cyclist or cyclist do this, be aware he may be about to manoeuvre.

B-7 Frequently (about every 4 to 5 seconds - more often in busy areas).

B-8 Mirrors - Signal - Manoeuvre.

B-9 They are vulnerable and difficult to see.

B-10 Pull in (where safe) to let other vehicles go past.

QUESTIONS MODULE C

C-1 What speed limit is usually indicated by the presence of street lights?

C-2 The Highway Code advises you to drive slowly in residential areas - Why is this?

C-3 What features might you find in some residential areas to slow traffic down?

C-4 What speed limit might be in force in some residential areas?

C-5 How does speed reduction affect pedestrians?

* * * * *

C-6 Why should you drive more slowly at night?

C-7 Apart from the speed limit in force, what decides the speed you should drive at?

C-8 Rule 57 of the Highway Code states that you should drive at a speed that will...

C-9 What gap should you allow when following behind other vehicles on the open road in good conditions?

C-10 In what way are the stopping distances of LGVs (large goods vehicles) and motorcycles different from cars?

ANSWERS MODULE C

C-1 30 mph, unless signs show other limits

C-2 There are pedestrians about and more vehicle movement around houses, shops etc.

C-3 Road humps (sleeping policemen), narrowings, traffic islands.

C-4 20 mph.

C-5 The lower the speed the less likely a pedestrian is to be killed in the event of a collision.

* * * *

C-6 You can't see as much and it's more difficult to see other road users, especially pedestrians and cyclists.

C-7 Road and traffic conditions, weather conditions. Distance you can see to be clear.

C-8 Allow you to stop well within the distance you can see to be clear.

C-9 A two-second gap (the two-second rule).

C-10 They need a greater distance to stop. (LGVs because of their extra weight - 7.5 to 38 tonnes - as opposed to a car [approx 1 tonne] and motorcycles because their braking systems are not as efficient as those on cars).

QUESTIONS MODULE D

D-1 You are overtaken by a vehicle that pulls in to the gap in front of you and reduces your available stopping distance. What should you do?

D-2 Before setting off on a journey in fog, what should you do?

D-3 What allowances would you make when considering a journey in fog?

D-4 Before driving in fog what would you check on your vehicle?

D-5 What should you do when driving in fog?

* * * * *

D-6 Driving along in clear conditions, you see a roadside signal displaying the word 'FOG'. Apart from 'FOG' what could you expect?

D-7 How should you prepare your vehicle for winter driving?

D-8 How would you drive in freezing or near freezing conditions?

D-9 The roads have just been gritted due to icy conditions. What should you be wary of?

D-10 When driving in snow what gear should you use? Why?

ANSWERS MODULE D

D-1 *Ease off and drop back to maintain a safe distance.*

D-2 *Decide if your journey is really necessary.*

D-3 *Allow yourself extra time. Set off earlier.*

D-4 *That all lights (including brake lights) are working and windscreen, windows and lights are clean.*

D-5 *Use dipped headlights, wipers and demisters, check your mirrors more often and drive more slowly. Don't hang on to the tail lights of the vehicle in front.*

* * * *

D-6 *Drifting smoke.*

D-7 *Make sure that the battery is well maintained and that there are appropriate anti-freeze agents in the radiator and screen-wash bottle.*

D-8 *With great care. Avoid harsh acceleration / braking / steering.*

D-9 *Surface conditions may change suddenly. Roads may be slippery.*

D-10 *The highest gear possible (This helps to avoid wheel spin).*

QUESTIONS MODULE E

E-1 When driving in snow, how should the controls be used?

E-2 What should you be aware of when snow-ploughs are operating?

E-3 When can you overtake a snow-plough?

E-4 When there are pedestrians about, how should you drive?

E-5 Name 6 situations which call for particular awareness of, and consideration towards, pedestrians?

* * * * *

E-6 Which type of pedestrians should you watch out for particularly? And why?

E-7 How could you identify blind / partially sighted pedestrians?

E-8 You see a pedestrian with a white stick with two red reflective bands. What does this mean?

E-9 What allowances should you make for pedestrians with any type of handicap?

E-10 Driving along, you see a flashing amber signal below a 'School' warning sign. What does this signal mean?

ANSWERS MODULE E

E-1 Very gently. This reduces the risk of skidding.

E-2 Snow may be being thrown out each side.

E-3 Don't; unless the lane you want to move into has been cleared of snow.

E-4 Carefully and slowly. Show consideration to, and for, pedestrians.

E-5 Crowded shopping streets; residential areas; bus and tram stops; parked milk floats and mobile shops; schools and colleges.

* * * * *

E-6 Children and old people. They find it more difficult to judge speed and distance.

E-7 They might be carrying a white stick and may also have a guide dog with them.

E-8 That the person is both blind and deaf.

E-9 Allow them more time to cross the road. Think of them and for them.

E-10 There may be children crossing the road ahead.

QUESTIONS MODULE F

F-1   When the amber lights are flashing near schools, what speed should you drive at?

F-2   When you pass a stationary bus showing a school bus sign, what should you be aware of?

F-3   When a school crossing patrol (Lollipop Lady/Man) shows the 'STOP - Children' sign, what must you do?

F-4   Who should you give way to at road junctions?

F-5   To reach a driveway, you need to cross a pavement. There are people on the pavement. What should you do?

* * * * *

F-6   Complete this statement from the Highway Code: 'Pavements are ...'

F-7   Where might you expect to find pedestrians walking in the road? What should you do?

F-8   On a road with no pavements, why should you take extra care on a left hand bend?

F-9   What should you do as you approach a zebra crossing?

F-10 When MUST you stop at a zebra crossing?

ANSWERS   MODULE F

F-1 Drive very slowly until you are well clear of the area.

F-2 Children may be getting on and off the bus, or may be in the vicinity of it.

F-3 You MUST stop.

F-4 Pedestrians who are already crossing the road into which you are turning.

F-5 Give way to them

* * * *

F-6 ... For people - not for vehicles.

F-7 In rural areas where there are no pavements, and roads may be narrow.

F-8 Pedestrians may be on your side of the road - slow down.

F-9 Check both sides for people waiting to cross.

F-10 You MUST stop when someone has a foot (or buggy/pram) on the crossing or is already walking across.

QUESTIONS MODULE G

G-1 Why would it be dangerous to wave pedestrians across at a zebra crossing?

G-2 If you see or hear an emergency vehicle, what should you do?

G-3 Looking in your mirror you notice a vehicle with a flashing green light. What does this mean? What would you do?

G-4 Driving along a fast main road you notice a vehicle in the distance with a flashing amber beacon. What does this mean and what would you do?

G-5 Give examples of vehicles which might use flashing amber beacons.

* * * * *

G-6 If the police wanted to stop your vehicle, how would they do it?

G-7 When requested to stop, what action MUST you take?

G-8 Complete this statement from the Highway Code : 'Give way to buses ... '

G-9 What should you be aware of around buses at bus stops?

G-10 You come up behind riders on horseback. What should you look out for, and be aware of?

ANSWERS MODULE G

G-1 Because other vehicles may be approaching and you could be waving pedestrians into danger.

G-2 Make room for them to pass - if necessary pull in to the side of the road and stop.

G-3 It's a doctor answering an emergency call - give way as soon as possible.

G-4 It's a slow moving vehicle or a broken down vehicle. You should check mirrors and plan your course.

G-5 Tractor / Road Gritter / Milk Float / Break-down Truck.

* * * * *

G-6 From behind they will flash their headlights and/or blue light and sound the siren/horn.

G-7 You MUST pull over and stop as soon as it is safe to do so and then switch off your engine.

G-8 ... Whenever you can do so safely - not just in towns.

G-9 People leaving the bus and crossing the road.

G-10 Signals from the riders. They may not move to the centre of the road prior to turning right.

QUESTIONS MODULE H

H-1 Where would you find passing places, and how would you use them?

H-2 What does the Highway Code state regarding diagonal white stripes/chevrons?

H-3 What is a 'crawler' lane, and how would you use it?

H-4 What is meant by 'lane discipline'?

H-5 Driving on a two-lane dual carriageway, when would you use the right-hand lane?

* * * *

H-6 How would you know when buses and trams have specific lanes?

H-7 When can you drive in a tram lane?

H-8 How would you overtake safely?

H-9 How much room should you give cyclists/motorcyclists when overtaking?

H-10 Why are road junctions hazardous? Who should you watch out for?

ANSWERS MODULE H

H-1 *On single track roads. Pull in to let others pass, or wait opposite a passing place.*

H-2 *If the area has a broken edge you may enter if you can see that it is safe to do so. If it has a solid edge you MUST NOT enter.*

H-3 *It is an extra lane, uphill, which you should use if you are driving a slow-moving vehicle.*

H-4 *Making sure you use the correct lane, and if changing lanes you must not force other road users to change speed or direction.*

H-5 *Only for overtaking or turning right.*

* * * *

H-6 *It would be indicated by road markings and signs.*

H-7 *Only when signs indicate you may do so.*

H-8 *Make sure the road is sufficiently clear ahead and behind - use the PSL MSM routine.*

H-9 *At least 6 ft / 2 m.*

H-10 *Because other road users are emerging and leaving one road to go into another : joining and crossing approaching traffic.*
Watch out for cyclists/motorcyclists and pedestrians.

QUESTIONS MODULE I

I-1 Which road users are most at risk at junctions?

I-2 Approaching a set of traffic lights you notice that they are not working. What do you do?

I-3 Before making a left turn, major-to-minor, what should you do?

I-4 You intend to turn left from a main road and you notice a rider on horseback just before the turning. What should you do?

I-5 To enter a road on the left you could need to cross a bus/cycle/tram lane. What should you do?

* * * * *

I-6 You're approaching a roundabout. When should you decide which exit you want?

I-7 If, on approach to a roundabout there are more than 3 lanes, which one would you choose?

I-8 If you were going 'full circle' at roundabout, how would you do it?

I-9 Which road users should you particularly look out for at a roundabout?

I-10 What advice does the Highway Code give regarding mini roundabouts?

ANSWERS MODULE I

I-1 Pedestrians, cyclists and motorcyclists.

I-2 Proceed with caution.

I-3 Check mirrors for traffic coming up on your left before you make the turn.

I-4 Hold back - don't overtake just before turning left.

I-5 Give way to any vehicles using it from either direction.

* * * * *

I-6 As early as possible.

I-7 Use the most appropriate lane on approach and through the roundabout.

I-8 Signal right on approach in the right hand lane. Continue to signal right until you have passed the exit before the one you want, then signal left.

I-9 Long vehicles, cyclists, motorcyclists and horse riders.

I-10 If possible, pass round the central marking, watch out for vehicles making a U-turn and for long vehicles.

QUESTIONS MODULE J

J-1 How far legally are you allowed to reverse your vehicle?

J-2 If you were parking in a driveway, would you reverse in or out of the driveway?

J-3 Which lights would you use 20 minutes after sunset?

J-4 Which lights would you use 40 minutes after sunset?

J-5 On which roads should you use headlights at night?

* * * * *

J-6 If your vehicle is fitted with dim/dip lights, when should you use them?

J-7 Only use front and/or rear fog lights when visibility is seriously reduced - what distance does the Highway Code advise?

J-8 What could you do to avoid being blinded by low (especially winter) sunshine?

J-9 If someone flashes their headlights at you, what should you never assume?

J-10 What special consideration, and why, should you give when parking next to, or behind a vehicle displaying a disabled person's badge?

ANSWERS MODULE J

J-1 Only as far as is necessary.

J-2 Where possible, reverse in and drive out; it's safer to do it this way because you have a better view of traffic when emerging.

J-3 Sidelights.

J-4 Headlights.

J-5 On all roads without street lighting. On roads where street lights are more than 185m apart. On lit motorways and on roads with a speed limit in excess of 50mph.

* * * *

J-6 In dull daytime weather and in built-up areas with good street lighting.

J-7 100 m (328 ft).

J-8 Make appropriate use of sun visor(s) and sun-glasses (keep a clean unscratched pair in the car). Plan ahead and if your route causes you to drive into the sun, prepare for it by using sun-glasses and/or sun visor(s). Don't look directly at the sun. Keep your windscreen clean and clear on the INSIDE as well as the outside. Because reflected glare can seriously affect your ability to see, take extra care when roads are wet.

J-9 That it is a signal to proceed.

J-10 Leave plenty of room so they can get in and out of their vehicle.

QUESTIONS MODULE K

K-1 What does the Highway Code state regarding parking on a red route ?

K-2 You are about to park at the kerbside when you notice that the kerb has been lowered. What does this mean ? What would you do ?

K-3 The Highway Code advises not to park within how many metres of a junction ?

K-4 You are approaching road works. What should you watch out for ?

K-5 At roadworks you notice speed limit signs. Are they advisory or compulsory?

* * * * *

K-6 If you saw the flashing lights of emergency vehicles moving very slowly or stopped, what could this mean ? What should you do ?

K-7 What does the Highway Code advise you to do when passing the scene of an accident ?

K-8 What should you do if you stop to give assistance at the scene of an accident ?

K-9 A vehicle marked with plain orange reflectorised plates is involved in an accident. What does this mean ?

K-10 At an accident, you are phoning the emergency services. What information should you give them ?

ANSWERS   MODULE   K

K-1 *You MUST NOT park.*

K-2 *It's been lowered to help wheelchair users. - Don't park there.*

K-3 *10 m / 32 ft.*

K-4 *Signs on the approach to, and at, road works.*

K-5 *Compulsory.*

* * * *

K-6 *There could have been an accident. Slow down and be prepared to stop.*

K-7 *Don't be distracted. Concentrate on your driving.*

K-8 *Warn other traffic. Arrange for the emergency services. Don't move injured people. Don't remove an injured motorcyclist's helmet. Stay at the scene until emergency services arrive.*

K-9 *It is carrying dangerous goods.*

K-10 *Location - Casualties. If the vehicles are carrying dangerous substances, give details of labels and other markings.*

QUESTIONS MODULE L

L-1 What is the hard shoulder on the motorway used for ?

L-2 Can you tell me the maximum speed limit on a dual carriageway for this particular vehicle ?

L-3 The acceleration lane on a motorway, where is it ?

L-4 On a 3 laned motorway what would you use the right-hand lane for ?

L-5 If your car broke down on the motorway what would you do ? How would you get help ?

* * * *

L-6 What is the procedure for joining a motorway ?

L-7 Which vehicles are not allowed to use the outside lane of a carriageway with three or more lanes?

L-8 Name 5 things you must not do on a motorway.

L-9 Travelling at 70 mph what is your overall stopping distance ?

L-10 What would you do if something fell from your car on the motorway ?

ANSWERS MODULE L

L-1 Only for accidents or emergencies.

L-2 70 mph (for cars / car derived vans / motorcycles) or specific limit for vehicle type.

L-3 It is a continuation of the slip road, and is used to build up speed to match that of traffic on the motorway.

L-4 Overtaking only.

L-5 Get it on to the hard shoulder and phone for help from one of the emergency telephones.

* * * * *

L-6 Approach on the slip road which will lead into the acceleration lane - use this lane to build up your speed to match that of the traffic on the motorway. Wait for a safe gap and join the inside lane. If there is no safe gap wait at the end of the acceleration lane.

L-7 Goods vehicles of more than 7.5 tonnes.
Any vehicle pulling a trailer.
Buses longer than 40ft / 12 m.

L-8 i) Exceed the speed limit.
ii) Reverse.
iii) Cross the central reservation.
iv) Walk on the carriageway.
v) Stop on the carriageway.

L-9 315ft / 96m.

L-10 Stop at the nearest emergency phone and inform the police. Do not try and retrieve it yourself.

QUESTIONS MODULE M

M-1 On a 3 lane motorway with no other traffic, what would your position be?

M-2 What colour, and where, are the studs on the motorway?

M-3 On a motorway how would you make other drivers aware of your presence, and why use that particular method?

M-4 How would you re-join the main carriageway of the motorway from the hard shoulder?

M-5 When driving in lanes what should your position be?

* * * *

M-6 What advice does the Highway Code give regarding speed when you're leaving a motorway?

M-7 On a 3 lane motorway when would you use the centre lane?

M-8 If you were driving along the motorway and you missed the exit you wanted, what would you do?

M-9 If you buy a cycle rack to fit on your car, what must you make sure of when you use it?

M-10 Who is not allowed to use a motorway?

ANSWERS MODULE M

M-1   In the left-hand lane.

M-2   Red   -   left-hand edge.
      Amber   -   right-hand edge.
      Green   -   marking acceleration and deceleration lanes.
      White   -   separate the lanes.

M-3   Flashing headlamps. The high road noise level may prevent them from hearing your horn.

M-4   Build up speed on the hard shoulder joining the inside lane of the motorway when there is a safe gap.

M-5   In the middle of the lane.

* * * * *

M-6   Check your speedometer.

M-7   If there were slower moving vehicles in the left-hand lane.

M-8   Drive on until you reach the next exit.

M-9   Make sure it is fitted properly. Any cycle(s) carried must be secure and the vehicle's lights and number plate must remain visible at all times.

M-10 Provisional licence holders, pedestrians, cyclists, horse riders, slow moving vehicles, agricultural vehicles, motor cycles under 50 cc.

QUESTIONS MODULE N

N-1 How do you turn right off a motorway?

N-2 Because traffic travels more quickly on a motorway, what should you do?

N-3 When joining a motorway, to whom must you give way?

N-4 If you are joining a motorway where the slip road becomes a lane on the motorway, what should you do?

N-5 What speed limit must you not exceed on the motorway?

* * * * *

N-6 Where is the only place you can park on a motorway?

N-7 What advice does the Highway Code give regarding breakdowns on the motorway?

N-8 If you breakdown on the motorway and cannot get your car on to the hard shoulder, what should you do?

N-9 If you have a disability, what should you do if you break down on the motorway?

N-10 On a motorway, if red lights (and/or a red X) flash above your lane, what would you do?

ANSWERS MODULE N

N-1 Leave by a slip road on your left. However, note that there are some lanes which lead directly off the motorway.

N-2 Think more quickly. Be aware of traffic much further ahead and behind.

N-3 You MUST give way to traffic already on the motorway.

N-4 Stay in that lane until it becomes part of the motorway.

N-5 You must not exceed the speed limit for your type of vehicle.

* * * * *

N-6 At a service area.

N-7 Pull on to the hard shoulder, as far to the left as possible. Switch on your hazard warning lights. Walk to an emergency telephone and give full details to the police.

N-8 Switch on hazard warning lights. Leave your vehicle only if you are sure you can safely get clear of the carriageway.

N-9 Switch on your hazard warning lights.
Stay in your vehicle with all doors locked.
Display a 'HELP' pennant.

N-10 You must not go beyond the signal in that lane.

QUESTIONS MODULE O

O-1 What does the Highway Code state regarding the flashing of headlamps?

O-2 If you were driving at 50 mph, how much space do you need to stop in?

O-3 Your car is fitted with two rear seat-belts. Your rear seat passengers are two adults and a child. Who should use the seat-belts?

O-4 What is the overall stopping distance on a good dry road at 40 mph?

O-5 Can you tell me what the flashing amber light means at a pelican crossing?

* * * *

O-6 What's the rule about a yellow box junction?

O-7 Your car starts to skid. What action would you take to help control it?

O-8 Normally you overtake on the right. Sometimes you can pass on the left. Can you name some of those times?

O-9 If you were turning right from a one-way street what position on the road should you be in?

O-10 If you were driving at night and the headlights from an oncoming vehicle were dazzling you, what would you do?

ANSWERS MODULE O

O-1 *Only use them to make others aware of your presence.*

O-2 *175 ft / 53 m.*

O-3 *The adults. Because of the heavier weight, an unrestrained adult would cause more injury in the event of an accident.*

O-4 *Overall stopping distance is 120 ft / 36 m.*

O-5 *Give way to pedestrians on the crossing.*

* * * *

O-6 *Do not enter unless your exit is clear. The only exception is if you want to turn right and are prevented from doing so by oncoming traffic.*

O-7 *Release the pressure on the footbrake, re-apply the pressure and then pump the brakes if necessary. If the back of the car skids to the right or left, steer into the skid.*

O-8 *i) When a vehicle is turning right and you can pass safely on the left.*
ii) When you are in the correct lane to turn left at a junction.
iii) When there are queues of slow moving traffic and vehicles in the lane on your right are moving more slowly than you are.
iv) When it is safe to do so in one-way streets.

O-9 *Well over to the right.*

O-10 *Slow down or stop. Also, look away from the source of dazzle.*

QUESTIONS MODULE P

P-1 Driving along you run into fog. What are the first few things you think about doing?

P-2 What do two solid white lines along the road mean?

P-3 At 30 mph on a dry road what is the overall stopping distance?

P-4 Between what times should you not sound your horn?

P-5 On a road with no parked vehicles, and where conditions allow, where would you position your car in relation to the kerb?

* * * * *

P-6 If you're driving at night on an unlit road, which of your lights would you use if you are following another car?

P-7 If you're turning right out of a narrow street where would you position your vehicle?

P-8 If you brake very hard and the car skids in a straight line what would you do?

P-9 What extra precautions would you take in the daytime if it was misty or foggy?

P-10 Tell me some of the places where you should not park.

ANSWERS MODULE P

P-1 *Slow down, keep a greater distance between yourself and the vehicle you are following. Use dipped headlights. Use windscreen wipers and demisters to keep the screen clean and clear. Don't hang on to the tail lights of the vehicle in front.*

P-2 *Do not cross or straddle.*

P-3 *75 ft / 23 m.*

P-4 *11.30pm and 7 am (in a built-up area).*

P-5 *3ft / 1m from the nearside kerb.*

* * * *

P-6 *Dipped headlights.*

P-7 *Well to the left.*

P-8 *Release, and then re-apply, the pressure on the footbrake. If the car still skids, repeat the procedure.*

P-9 *Use dipped headlights, slow down. Keep a greater distance from the vehicle in front.*

P-10 *Opposite traffic islands, on or near the brow of a hill, where it would endanger or inconvenience other road users, where it would hide a traffic sign.*

QUESTIONS MODULE Q

Q-1 If you're on a country road what special dangers would you be aware of?

Q-2 What precautions should you take when driving in the rain?

Q-3 When would you be allowed to wait in a box junction?

Q-4 On a right-hand bend what should your position be?

Q-5 What would you do if the rear of the car skidded to the right?

* * * *

Q-6 How does drinking alcohol affect driving ability?

Q-7 What should you be on the look out for just before you enter your car?

Q-8 Are you allowed to drive in bus lanes at all?

Q-9 What is the last thing you check before getting out of your car?

Q-10 How far away from your vehicle should you place a red warning sign?

ANSWERS   MODULE Q

Q-1 *Pedestrians and/or animals in the road, agricultural vehicles.*

Q-2 *Slow down, keep a greater distance from the vehicle in front, use wipers and demisters.*

Q-3 *If you were turning right and are prevented from doing so only by oncoming traffic.*

Q-4 *Well to the left.*

Q-5 *Steer to the right.*

* * * *

Q-6 *It reduces co-ordination, increases reaction time and impairs judgement of speed, distance and risk.*

Q-7 *Cyclists, vehicles, and passing pedestrians.*

Q-8 *Yes, outside the bus lane operating times.*

Q-9 *Check over your shoulder before opening the door.*

Q-10 *50 metres. On a motorway 150 m.*

QUESTIONS MODULE R

R-1 What is the national speed limit on a single carriageway road?

R-2 How should you drive when passing animals?

R-3 What does a parked vehicle in the road tell you?

R-4 When in charge of a vehicle, when may you remove your seat belt?

R-5 When driving on ice how should you use the controls of the car?

* * * * *

R-6 After overtaking, what must you never do?

R-7 When turning into a road junction to whom must you give way?

R-8 Before you drive, you must be able to read a number plate at what distance?

R-9 After driving through a stretch of flooded road what should you do?

R-10 Where may you park at night without lights?

ANSWERS MODULE R

R-1 *60 mph.*

R-2 *Slowly, allow them plenty of room. Do not rev the engine or sound the horn.*

R-3 *Road narrows.*

R-4 *When carrying out a manoeuvre that involves reversing.*

R-5 *Very delicately and smoothly.*

* * * * *

R-6 *Cut in, slow down or stop.*

R-7 *Pedestrians crossing the road into which you are turning.*

R-8 *20.5 m (about 67 ft). Note, however, this is only 50% of perfect vision and is the absolute minimum standard.*

R-9 *Dry your brakes by driving very slowly with your left foot pressing lightly on the brake pedal.*

R-10 *On the road (if in a 30 mph limit) facing the direction of traffic flow. In a recognised parking place.*

QUESTIONS MODULE S

S-1 What must you always do before changing direction?

S-2 Driving down a road with parked vehicles what dangers would you be aware of?

S-3 What should you never leave in an unventilated car when you park?

S-4 In a built-up area where you have street lighting, what would the speed limit usually be?

S-5 At 60 mph in good conditions what is your overall stopping distance?

* * * * *

S-6 How does a bus lane work?

S-7 If you're approaching a roundabout and you want to follow the road directly ahead, which is usually the second exit, how would you do that?

S-8 What does the overall stopping distance consist of?

S-9 If you were skidding to the left, which way would you steer?

S-10 When can hazard warning lights be used on a moving vehicle?

ANSWERS MODULE S

S-1 Check the mirrors.

S-2 Vehicle occupants may open doors.
Pedestrians (especially children) may walk out between vehicles.
Vehicles could drive off without signalling.

S-3 Children or animals.

S-4 30 mph. There are exceptions - usually a higher limit is indicated by signs on the lampposts

S-5 240 ft / 73 m.

* * * * *

S-6 A bus lane operates between certain hours, which will be indicated by a time plate. Outside those hours you may drive in the bus lane.

S-7 Approach in the left hand lane, if the roundabout is clear drive around the roundabout in the left-hand lane and indicate left at exit before the one you want.

S-8 Thinking distance and braking distance.

S-9 To the left.

S-10 To warn others briefly if you have to slow down quickly on a motorway or unrestricted (70 mph) dual-carriageway.

QUESTIONS MODULE T

T-1 What is the maximum speed limit on a dual carriageway?

T-2 What do white diagonal stripes in the road mean?

T-3 What is the 2-second rule?

T-4 When should you give way to a bus?

T-5 What type of signs are triangular?

* * * * *

T-6 You should only use rear fog lights if visibility is reduced to what?

T-7 At a junction in fog what could you do to help locate other traffic?

T-8 How could you tell if a pedestrian is blind and deaf?

T-9 Who should you be particularly aware of at zebra crossings?

T-10 If a zebra crossing is divided by a central island unit what does this mean?

ANSWERS MODULE T

T-1 70 mph.

T-2 They keep traffic streams apart which may be a danger to each other.

T-3 To ensure you leave a safe gap between your vehicle and the vehicle you are following, use the 2-second rule :- when the vehicle in front passes a fixed point (bridge, telegraph pole) count 2 seconds by saying slowly "1 second, 2 seconds". If you can say this before you reach the fixed point you are following at a safe distance.

T-4 When it is signalling to move away, and you can safely give way to it.

T-5 Warning signs.

* * * *

T-6 Less than 100 m.

T-7 Open your windows and listen.

T-8 He would carry a white stick with 2 reflective bands.

T-9 Old people, young children and people with prams.

T-10 Treat each half as a separate crossing.

QUESTIONS MODULE U

U-1 If you are dazzled by a vehicle behind you, what should you do ?

U-2 When turning right at a junction where there is an oncoming vehicle also turning right, you would generally pass offside-to-offside. Can you tell me when you would pass nearside-to-nearside ?

U-3 When may you cross a solid white line along the road ?

U-4 Before turning left, what must you always check for ?

U-5 What does the Highway Code state regarding your vehicle's condition ?

* * * * *

U-6 What items on your vehicle should you check regularly ?

U-7 Before you drive, what should you ensure are adjusted correctly ?

U-8 A motorcycle rider and pillion passenger must do what ?

U-9 Before you set off on a drive you feel tired or ill. What does the Highway Code state regarding driving if you are in such a condition ?

U-10 If you feel tired whilst driving, what should you do ?

ANSWERS MODULE U

U-1 *Move your head slightly to avoid the dazzle. Do not adjust your mirror as you may not remember to re-set it.*

U-2 *If the layout of the junction or traffic situation makes offside to offside crossing impractical.*
If indicated by road markings or the position of the other vehicle.

U-3 *To enter premises or a side road, to avoid a stationary obstruction, or when ordered to do so by a police officer or traffic warden.*

U-4 *To make sure that a cyclist or motor-cyclist is not coming up from behind on your left.*

U-5 *It MUST be roadworthy.*

* * * * *

U-6 *Windscreen washers/wipers, tyres, exhaust system, lights, brakes.*

U-7 *Seat, seat-belt, mirrors, head restraint.*

U-8 *MUST wear an approved safety helmet which MUST be fastened securely.*

U-9 *DO NOT DRIVE.*

U-10 *Find a safe place to stop and rest.*

QUESTIONS MODULE V

V-1 What would you do if red lights were flashing on the central reservation or on a slip road?

V-2 If amber lights are flashing on a motorway, what should you do?

V-3 What is the cause of 95% of road traffic accidents?

V-4 How often should you check your tyre pressures?

V-5 At an automatic half-barrier level crossing what would you do if the amber lights flash and the alarm starts when you have crossed the white line?

* * * * *

V-6 Can you drive in a road or lane reserved for trams?

V-7 What do yellow zig-zag lines mean outside a school?

V-8 How would you inform someone controlling traffic that you wanted to go straight ahead?

V-9 Which of the traffic light colours means stop?

V-10 When there are double white lines along the road and the line nearest to you is continuous, what does this mean?

ANSWERS MODULE V

V-1 You MUST not go beyond the red lights in any lane.

V-2 Reduce your speed, look out for the danger and do not increase your speed until you pass a signal which is not flashing.

V-3 Driver error.

V-4 Every week.

V-5 Keep going.

* * * * *

V-6 No. You must not enter such a road or lane.

V-7 Keep entrance clear of stationary vehicles, even if picking up or setting down children.

V-8 Hold your left hand up to the windscreen, palm facing forwards.

V-9 They all do, except green which means you may go only if the junction is clear.

V-10 Do not cross or straddle the solid line nearest to you.

QUESTIONS MODULE W

W-1 Zig-zag lines at pedestrian crossings. What do they mean?

W-2 What is the one situation when an arm signal should be used?

W-3 Which two colours does a traffic light show at the same time?

W-4 Driving along you see diamond shaped signs. Who are these for?

W-5 If you break down or have an accident on a level crossing, what should you do?

* * * * *

W-6 What fluid levels should you check regularly?

W-7 What are the tyre pressures for this particular vehicle?

W-8 If you don't know what the tyre pressures are for the vehicle you are driving, how could you find out?

W-9 What advice does the Highway Code give regarding open level crossings?

W-10 When should you give way to trams?

ANSWERS   MODULE W

W-1 *No parking, no stopping, no overtaking.*

W-2 *If you are the first vehicle on approach to a zebra crossing, and you have time.*

W-3 *Red and amber.*

W-4 *Tram drivers.*

W-5 *Get everyone out of the vehicle and clear of the crossing. If there is a railway telephone use it to tell the signal operator. If time allows, move the vehicle clear of the crossing.*

* * * * *

W-6 *Windscreen washer reservoir (front and rear), oil, brake fluid reservoir, radiator, battery.*

W-7 *Front :*   *Rear :*

  Insert correct details for the vehicle(s) that you drive.

W-8 *Look in the vehicle's handbook or on a wall chart at a filling station.*

W-9 *Look both ways, listen and make sure there is no train coming !*

W-10 *Always. Don't try to race them or overtake them.*

QUESTIONS MODULE X

X-1 When you approach traffic lights, the lights are showing amber, what does this mean?

X-2 Demonstrate the arm signal for slowing down.

X-3 When a tram is approaching a stop, who should you look out for?

X-4 What type of fuel does this vehicle run on?

X-5 What's the arm signal for turning left?

* * * * *

X-6 In the traffic light sequence, which light follows immediately after the amber light on its own?

X-7 What does the Highway Code state regarding parking your vehicle and trams?

X-8 If you are involved in an accident which causes damage or injury, what must you do?

X-9 Can the police stop your vehicle at any time?

X-10 If the police stop you, what documents may they require you to produce?

ANSWERS MODULE X

X-1 Stop, unless you have already crossed the white stop line, or to pull up would cause an accident.

X-2 Right arm outstretched, palm facing downwards, move the arm up and down.

X-3 Look out for pedestrians, especially children, running to catch it.

X-4 Leaded/unleaded petrol, diesel, derv, petrol/oil 2-stroke mix. It will be **one** of these, not a combination.

X-5 Right arm outstretched, palm towards the ground, move your arm round in a circle (anti-clockwise).

* * * * *

X-6 Red.

X-7 You must NOT park your vehicle where it would get in the way of trams, or force other drivers to do so.

X-8 Stop. Give your details to anyone having reasonable grounds for requiring them; if not, report the accident as soon as reasonably practicable, and in any case within 24 hours.

X-9 Yes.

X-10 Driving licence, insurance certificate, MOT certificate (if applicable).

QUESTIONS MODULE Y

Y-1 As a driver of a motor vehicle, what valid documents must you have?

Y-2 What is the minimum legal tread depth on tyres?

Y-3 If you develop a health condition that is likely to affect your driving; what must you do?

Y-4 When a school crossing patrol person (Lollipop Man/Lady) shows a 'STOP- Children' sign, what must you do?

Y-5 What is the cause of 95% of road traffic accidents?

* * * * *

Y-6 If you have been stopped by the police and don't have the required documents with you, you must produce them at a police station of your choice within how many days?

Y-7 Driving along you unavoidably run over a cat. Must you report the accident?

Y-8 How many penalty points would result in disqualification?

Y-9 Which driving offences would result in disqualification and an order to take an extended re-test?

Y-10 Can the courts order a re-test for any road traffic offence which carries penalty points?

ANSWERS MODULE Y

Y-1 Driving licence, tax disc, third party insurance, MOT certificate (if applicable).

Y-2 1.6mm across the central three quarters of the width, with visible tread across the remainder of the width.

Y-3 Report it to the licensing authority (DVLA, Swansea).

Y-4 Stop. Don't move away until the sign is removed.

Y-5 Driver error.

* * * * *

Y-6 Seven days.

Y-7 Legally no, but if possible (depending on circumstances) it would be humane to move the body to the side of the road and call a vet if the animal were still alive.

Y-8 12 or more within a 3-year period.

Y-9 Causing death by dangerous driving. Dangerous driving.

Y-10 Yes.

QUESTIONS MODULE Z

Z-1 You come upon an accident where a motorcyclist is lying injured on the ground. What should you do?

Z-2 If there are casualties in vehicles what should you do?

Z-3 What should casualties be given to drink - A hot drink, a cold drink, or nothing at all?

Z-4 What advice does the Highway Code give regarding extra security for your vehicle?

Z-5 How can you help to prevent your car being stolen?

* * * * *

Z-6 Protection markers are required where a load or equipment overhangs, front or rear, by how much?

Z-7 What does the sign HR (black HR on yellow background) mean?

Z-8 You are driving a car towing a caravan. What are the speed limits on single/dual carriageways and motorways.?

Z-9 You hire a transit van to move some belongings. What is the maximum speed you can do on a dual-carriageway?

Z-10 Why are Stop signs octagonal (eight-sided)?

ANSWERS MODULE Z

Z-1 Do NOT remove the motor-cyclist's helmet unless you have had proper medical training. Further serious injury could result from its removal.

Z-2 As long as their airways are clear and there is no danger from further collisions or fire, leave them in their seats. Removing them could cause further serious injury.

Z-3 Nothing. There may be unseen internal injuries and any fluids could make them worse.

Z-4 Fit an anti-theft device such as an alarm or immobiliser.

Z-5 Lock the car, close all windows (and sunroof) completely, take valuables with you, don't leave anything on show that would tempt thieves.

* * * * *

Z-6 2 m.

Z-7 Holiday route.

Z-8 Single carriageway 50 mph, dual-carriageway 60 mph, motorway 60 mph.

Z-9 60 mph.

Z-10 So that the shape and meaning can be readily identified even if the face of the sign is obscured (by snow for example). It is the only sign which has this shape and is internationally recognised as a stop sign.

ONE GOLDEN RULE

VARIETY IS THE SPACE OF LIFE

Although all of the rules in the Highway Code are to be followed and adhered to, there is one - Rule 57, which you should learn by heart and remember for the rest of your driving career, and live by it.

RULE 57   Drive at a speed that will allow you to stop well within the distance you can see to be clear.

Leave enough space between you and the vehicle in front so that you can pull up safely if it suddenly slows down or stops.

The safe rule is never to get closer than the overall stopping distance.

In good conditions on roads carrying fast traffic, a two-second gap may be sufficient. The gap should be at least doubled on wet roads and increased further on icy roads.

Large vehicles and motorcycles need more time to stop than cars.

Drop back if someone overtakes and pulls into the gap in front of you.

Driver error is the cause of 95% of road traffic accidents.

On average, 5,000 people are killed in road accidents each year. 4,750 of these deaths could have been prevented.

THINK ABOUT IT

The sign says stop, the line says not. What should you do at this junction? *Answer on p 87*

* * * * *

You have been warned !

SECTION 10

TRAFFIC SIGNS

IDENTIFY THE FOLLOWING TRAFFIC SIGNS

1.................

2.................

3.................

4.................

5.................

6.................

7.................

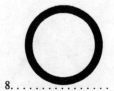
8.................

TRAFFIC SIGNS

IDENTIFY THE FOLLOWING TRAFFIC SIGNS

9.................

10.................

11.................

12.................

13.................

14.................

15.................

16.................

TRAFFIC SIGNS

IDENTIFY THE FOLLOWING TRAFFIC SIGNS

17. ....

18. ....

19. ....

20. ....

21. ....

22. ....

23. ....

24. ....

TRAFFIC SIGNS

IDENTIFY THE FOLLOWING TRAFFIC SIGNS

25. ....

26. ....

27. ....

28. ....

29. ....

30. ....

31. ....

32. ....

TRAFFIC SIGNS

IDENTIFY THE FOLLOWING TRAFFIC SIGNS

33. ....

34. ....

35. ....

36. ....

37. ....

38. ....

39. ....

40. ....

What do the signs mean?
What should you do?   *Answer on p 87*

* * * * *

Don't argue with this!

SECTION 11

ANSWERS TO TRAFFIC SIGNS

IDENTIFICATION OF TRAFFIC SIGNS

1. Change to opposite carriageway (may be reversed)
2. Road humps ahead
3. Mini-roundabout
4. No pedestrians
5. No U-turns
6. Dual carriageway ends
7. Road works ahead
8. No vehicles

* * * * *

9. Double bend first to left
10. Two way traffic ahead across a one way street
11. Vehicles may pass either side to reach the same destination
12. No entry
13. Waiting restrictions apply
14. Road narrows on right
15. Side road
16. No motor vehicles

ANSWERS TO TRAFFIC SIGNS

IDENTIFICATION OF TRAFFIC SIGNS

17. Hump back bridge
18. Temporary maximum speed limit on motorway
19. Roundabout
20. Ahead only
21. Lane(s) closed ahead
22. Cattle
23. Uneven road surface
24. Traffic signals ahead (may also be used to warn of pelican crossing)

* * * * *

25. Children going to and from school
26. No overtaking
27. Manually operated roadworks sign : Stop
28. Turn left ahead
29. No stopping ('clearway')
30. End of restriction (on motorway)
31. No left turn
32. Give priority to vehicles from the opposite direction

ANSWERS TO TRAFFIC SIGNS

IDENTIFICATION OF TRAFFIC SIGNS

33. No vehicles over width shown

34. Keep left (right if symbol reversed)

35. Leave motorway at next exit

36. Bend to right

37. Road narrows on both sides

38. Turn left

39. School crossing patrol

40. Entry to 20 mph zone

* * * * *

ANSWERS TO TRAFFIC SIGNS on Pages 78 and 84

Page 78   You must stop. The sign has the authority; in the event of the lines being covered up (snow, etc) the sign would still be recognisable.

Page 84   The top sign means vehicles over 6'6" width are prohibited. The lower sign indicates that you must give priority to vehicles from the opposite direction.

Both of these signs apply to the road to the left. This is indicated by the arrow underneath the signs.

You should, therefore, emerge very carefully - not everybody will have read or understood these signs!

SECTION 12

AND FINALLY ...

When you pass your driving test, consider it as the next step in your driving career.

The only thing that passing the test tells you is that on that day, at that time, your driving reached a minimum basic standard. That's all.

As drivers, we are all only as good (or bad) as the last time we drove.

EXPERIENCE COUNTS

As you become more experienced you will encounter different traffic situations, drivers, roads etc. Use this experience to plan your driving; judge what might possibly happen and make allowances for it. There is a well-known 'law' which states that 'if something can happen, it will!' Forewarned is forearmed.

ADVANCE AND BE RECOGNISED

Taking an advanced driving test is a natural progression from taking the basic driving test; it's like moving on to 'A' level after 'O' level.

If you don't fancy the idea of advanced driving, consider at least having your driving assessed on a regular basis (say every 3 years) by an ADI (Approved Driving Instructor).

STANDARDS

WITHIN THE

DRIVER TRAINING INDUSTRY

To teach driving for payment, a person must be registered as an ADI (Approved Driving Instructor)

To qualify as an ADI, there are three stringent tests that need to be passed: 1. A written test; 2. An eyesight test and an hour long advanced driving test; 3. An instructional ability test. Having passed these three tests, the person's name must then be entered on the Register of Approved Driving Instructors. A certificate of registration (bearing name and photograph) is then issued and this must be displayed in the windscreen when tuition is being given.

CHECK TESTS

One condition of an ADI's registration is that he/she must submit themselves for regular 'check tests' to ensure that they are of 'continuing ability and fitness to give instruction'. These check tests are conducted by a supervising examiner (ADI) who 'sits in' on a lesson and monitors the ADI's performance.

GRADED INSTRUCTORS

At the end of a check test the supervising examiner and the ADI will discuss the session and a grading will be awarded to the ADI. The grades are 1-6 and reflect various aspects of the ADI's tuition with six being the highest. Those ADI's who get a grade of 1, 2, or 3 are required to undergo further check tests(s) until they either achieve an acceptable standard, or are removed from the register altogether.

WHO TRAINS THE ADIs?

ADIs are trained by individuals or establishments which have set themselves up as trainers.

There are no compulsory qualifications required to do this. However, as from 1992, trainers have been able to apply to be entered in a new directory.

This is known as the ADITE (Approved Driving Instructor Training Establishment) directory.

Before becoming approved, training establishments/individuals undergo a one day inspection of training methods and premises by a supervising examiner (ADI), appointed by the DSA (Driving Standard Agency).

A FULL REPORT

The inspector will then file a comprehensive report concerning all aspects of the trainer's methods, facilities and courses available. The report will then be sent to the ADITE management committee for their consideration and approval (or not, as the case may be). This committee comprises representatives from various associations of driving instructors.

ONLY APPROVED ESTABLISHMENTS

Once an establishment has been approved, its name will be entered into the directory, and sent out in a list contained in the information package sent out to prospective driving instructors (PDIs). Essentially, this new directory will help PDIs by identifying *only* those training establishments which have been inspected, found to meet certain stringent criteria and have been approved. All such approved trainers must display, and adhere to, an agreed code of practice.

BETTER FOR EVERYBODY

This system should ensure that there are certain standards and that those being trained as instructors will get effective, accurate and value for money tuition. In turn those new instructors will teach 'L' drivers effectively, accurately and safely.

FEWER ACCIDENTS

So, where does this all lead? Better educated instructors lead to better educated drivers, which leads to a reduction in the number of road accidents.

The most recent figures available show that the road accident death toll for 1991 was 4,568.

If you consider that 95% of all accidents are caused by driver error then 4,339 of those deaths could have been prevented.

EFFECTIVE INSTRUCTOR AND DRIVER EDUCATION
MUST BE THE ANSWER

ADI PART III : ESSENTIAL INFORMATION

The Test of Ability to Instruct

by

Brian M Stratton DTp ADI

Revised, updated and enlarged edition published 1992

Clear and concise, this publication is a must for anyone involved in Part III training, and will be an invaluable addition to any training course.

Reflecting what candidates really want to know about the Part III, this book sets out, in a straightforward manner, all aspects of the Part III, including:

* How the Part III exam is marked
* Using the Q/A Technique
* 'How will I know when I'm ready?'
* Choosing a trainer

By giving a clear perspective, or over-view, of this most exacting exam, candidates will have a sharper picture of the content and methods involved.

Some comments on this publication:

"... I would like to take this opportunity to thank you for all the valuable information and advice contained in your Part III book ... I do not think I would have passed first time without your help"
A Millington, newly qualified ADI

"... I would have no hesitancy in recommending your books to any potential ADI or existing ADI"
A Cirket, newly qualified ADI

"... I think you have done a very good job of laying the information out in the publication ... I now supply it as part of my Part 3 course to trainees ... Keep up the great work for the profession".
A White, Diamond Advanced Instructor, ADITE Registered Trainer

"... The information contained within your 'Essential Information' document is impeccably presented and gives a clear understanding as to what is required ... No praise for your information notes can really express my heartfelt gratitude"
M Woods, newly qualified ADI

"... Introductions, points, errors, hints - very good ... excellent confidence booster"
I Hopkins, PDI

For details of availability and price, please contact the publishers:
First Time Publishing Tel: (0860) 260720

HILL START BLUES

A New Driving Manual for the 1990s

by

Brian M Stratton DTp ADI

ISBN : 0-9514415-6-6   £6.50

(Revised Edition published August 1991)

This book has been designed to help learner drivers by explaining driving tasks in a clear and logical manner. The first chapter deals with driving lessons (what the pupil might expect, etc) and subsequent chapters cover the various driving tasks in a similar way to the training programme of an ADI.

Published in January 1990, **Hill Start Blues** has been well received by ADIs and pupils. All have commented on the clarity of the unique lay-out, and the straightforward way in which information is presented.

The new syllabus for learner drivers is set out in detail. There is also an explanation of the changes to the Driving Test in 1991 which will harmonise the GB test with that of other EEC member countries.

Comments regarding this book:

"... It's a book that talks to you, it's not jargon or too technical"
J Allsop, Pupil

"... A definitive Driving Manual, a must for all ADIs and PDIs ..."
M Collins, ADI

"... All learner drivers (and ADIs!) could profit from this book ..."
The Times

"... A driving manual with a difference ... takes into account changes to the driving test in the 1990s"
Auto Express

"... Particularly liked the fault correctors and felt the road markings were more clearly explained than in the Highway Code"
RoSPA (Royal Society for the Prevention of Accidents)

"... And, an excellent paragraph called 'Fault Correctors', from which we can all learn something"
IAM (Institute of Advanced Motorists)

"... The book is well paragraphed and covers the full information needed for learning and taking the driving test"
RAC

"... Congratulate you on the content and format ..... In a class of its own where driving manuals are concerned ..."
E Daly, PDI

This book may be purchased direct from the publishers. Telephone for details: **First Time Publishing : (0860) 260720**

THE DRIVING TEST - *Graphic Traffic Version*

by

Brian M Stratton DTp ADI

ISBN 0-9514415-7-4   £5.95

(First Published September 1991)

This title supersedes **'The Driving Test : Essential Information'** a text book first published in 1989.

The Driving Test : *Graphic Traffic Version* is unique amongst driving reference books in that it consists almost entirely of illustrations, many of which have a humorous slant.

Both ADIs and pupils like the lay-out and the way in which the information is presented.

Extensively researched, the opinions of ADIs and learner drivers were sought at all stages during the writing of the book.

How the 'L' test is marked, Terminology and Highway Code Questions and Answers are just a few of the chapter headings in this useful reference book.

Some comments regarding this book:

"... Probably the best start-up book you can find - it's great fun and great advice"
Auto Express

"... Many hints to help you pass first time"
Institute of Advanced Motorists

"... An interesting and fresh approach to the subject"
RoSPA (Royal Society for the Prevention of Accidents)

"... Amusing cartoons ... get points across effectively and raises a smile in the process"
Driving Magazine

"... The summaries at the end of each chapter are ideal for quick reference"
J Follows, Learner driver

"... A runaway success with my pupils, nobody can put it down!"
L Lewis, Approved Driving Instructor

"... It's funny and the information is sound.   And it's so much easier to read"
J Voss, Newly qualified driver

This book may be purchased direct from the publishers.
Telephone for details: **First Time Publishing:** (0860) 260720

ADI PART III & INSTRUCTOR'S GUIDE

by

Brian M Stratton DTp ADI

(First Published 1992)

Available as individual volumes or as a complete set (20 volumes)

These guides are complementary to ADI Part III : 'Essential Information' and provide detailed and specific information on each topic of the Part III, at Beginner, Partly Trained and Trained standard.

Although primarily designed to help candidates with the Part III, they will serve as a reference book throughout an ADI's career and can be used as a memory-jogger to check tests and as a general source of information whenever needed.

Some comments regarding this publication :

" ... I like the way it's laid out as regards exactly what to say and do ... assessment of faults ... generally very very good indeed."
M Scott, PDI

" ... The biggest help is the word-for-word briefings - I was working blind before ... They cover all the points ... How to put it together and how to present it."
S Byford, PDI

"... Excellent, just what I've been looking for ... very pleased with them ... Well worth the money."
P Hebblethwaite, PDI

" ... Helping me considerably ... Absolute confidence builder ... Nothing skimped or corners cut ... Everything there ... I was in the wilderness before, now you've given me a map!"
D Ronaldson, PDI

" ... These guides definitely helped me to pass my Part III. The diagrams have all the necessary information on them and help you through the subjects."
J Laing, ADI

For more details of these guides, please contact the publisher :
First Time Publishing: (0860) 260720

INSTRUCTOR and DRIVER TRAINING
INSTRUCTOR and DRIVER TRAINING
INSTRUCTOR and DRIVER TRAINING
INSTRUCTOR and DRIVER TRAINING
INSTRUCTOR and DRIVER TRAINING
INSTRUCTOR and DRIVER TRAINING
INSTRUCTOR and DRIVER TRAINING
INSTRUCTOR and DRIVER TRAINING
INSTRUCTOR and DRIVER TRAINING
INSTRUCTOR and DRIVER TRAINING
INSTRUCTOR and DRIVER TRAINING
INSTRUCTOR and DRIVER TRAINING
INSTRUCTOR and DRIVER TRAINING
INSTRUCTOR and DRIVER TRAINING

* *ADI Training Courses* *

* *Confidential Check Test Guidance* *

* *Assessments for the Cardington 'Special' Driving Test* *

Purchasers of this publication qualify for discount on any of the above. Please contact Brian M Stratton, on the number shown below, for further details.

TELEPHONE (0860) 260720